WHAT WE SHOULD KNOW ABOUT WATERS THAT FLOW!

Human Environment

- We need the physical environment to meet our needs
- We obtain food from plants and animals, wood from trees and water from rivers
- Materials we use from the physical environment is called natural resources
- When we extract and use these natural resources we often change the physical environment

Rose Marie B. Wolford-Zabala

To the reader ';and read to;' of this book:
There's always a reason a story is written,
to find the "why" just back and listen.
Water, water everywhere! NO, NOT TRUE becoming rare!
Global waters we all share are growing scarcer--SO BEWARE:
This menacing, deadening, threatening of lands becoming bare!
Water levels are going down each and every day,
this needs our prompt attention-WE NO LONGER CAN DELAY!
Unless we try to find the WHYS-eyes opening by focusing
by charting and starting the 'reckoning of this unquestioning beckoning';
this unsettling, leveling of waters decreasing,
our world be in JEOPARDY~SCARCITY increasing.
We ALL must give attention, we MUST MORE than simply mention
With binding minds together find UNITED FORCE PREVENTION!

Special thanks to Dolores (BB..), French & Vilma Wolford

Carrie Campbell; Counselor/SUDCCII #6873, Palmdale, Ca. U.S.A.
Dr. Billy E. Lawrence, Theologian & Author; Colorado Springs, CO. U.S.A.
Susan Markebjer - Safström; Behavioral/Counselor: Youth & Early Intervention, Stockholm Vallentuna, Sweden
© Registration Number / Date:TXu001992006 / 2015-10-30

This book is dedicated to Mr. SEAMONSTER and his grandfather Juan de la Cruz Bona Ventura: For their struggle to make things right, with all their might, to end the blight of unrest-distress. Helping to bring peace so troubled waters cease!

Faithful Father
Firmly Foretelling
Forever Fondly Forgiving
Fellows, Friends and Family
But let justice rundown like water
And righteousness like a mighty stream
Amos
5:24

Rose Marie Wolford-Zabala is a Cognitive Behaviorist, Special Educator/Advocate and former Children Social Worker. Her experiences moved her to write the rhyming book series addressing current social-emotional, familial and environmental issues today experienced by families. At this point, the eighth in the series.

Rosey so Nosey and Roughest the Toughest series addressing:

Domestic Violence	Reading	Bullying
Divorce/Separation	Teamwork	Pollution Climate Change
Family and Global diversity	Hazards of Lying	Substance Abuse
Fear/Anxiety/Secrets	Stealing/Wrong conclusions	Water Conservation
Siblings fighting/reconciliation	Friends fighting and forgiving	Health Crisis/Safety

Rosey loved to explore and learn of "new things"
Not afraid of things different, while others did cling
To the same, not arrange, often shying from change
Due to fear of UNKNOWN and things out of range.
Rosey searched for the new and embraced this new treasure
Shared "discovers" with others for it brought her much pleasure
Roughest, her best friend helps find a way
Yes, they CAN find solutions and help save the day!
Let's all read together find what happened before
For the problem in store and we'll learn all the more
How Rosey and Roughest push problems away
Let's find out! All about…see what happened that day.

Rosey woke up with a **terrible fright!**
She tossed and she turned
ALL THROUGH THE NIGHT...
for the *dream* that she *dreamt*

WAS A HORRIBLE SIGHT!

That IT really could happen...

a CHANCE that IT might?

She talked to her folks, both mom and dad.
She told them about the dream that she had,

how it not only SCARED her...

made her SAD, made her MAD!
With confusion and doubt
she told them about

the PLIGHT and

THE FRIGHT

that she'd dreamt through the night.

They listened so carefully then stopped to pause
to share with their daughter
IT should and IT ought to...
THE REASON, THE CAUSE!
For Rosey had dreamt of the waters on earth,
how they no longer *flowed*
and watered NEW BIRTH.
Her parents told Rosey she'd
dreamt of a DROUGHT!
Proceeding with heed and needing **NO DOUBT**
to tell her just HOW THIS ALL comes about!
They both spoke with her
SO she'd KNOW for sure:
"When people aren't careful...
waters decrease, the flowing may cease!
Many folks do not care for the waters we share!
Yet, ALL of the world
DEPENDS on ITS FLOW
To nourish the earth and help
ALL THINGS GROW!"

Rosey was curious and asked,
"WHY'S IT SO?
Why people don't care,
why people don't share:
Don't they know it's important for
ALL things to grow?
Why don't they CARE about
WATERS THAT FLOW?
What will happen to things that
need water to thrive?
If the waters don't flow
how can they stay alive?
Will the waters that rain continually give
what's vitally needed so ALL things CAN live?
When the rain DOES NOT COME...

WHERE'S the WATER COME FROM?"

ISIAH 55:10

Sighing, replying, far from glad-Mom and Dad said sad,
"Earth has natural forces helping ITS RESOURCES to replenish and sustain.
Yet, IT remains to disarrange- threatening WASTE AND CLIMATE CHANGE!

Glaciers, rivers, waterfalls, streams or oceans, bays and seas; All these naturally providing food, world energy, habitats and world housing. In reality, ARE ACTUALLY in danger from all business profit gains; Despite the FRIGHT AND PLIGHT of pained and gained ecological strains! Consumer humans abuse, misuse, continuing on to infuse; un-natural forces to RESOURCES USED for INDUSTRY, ECONOMY- YET, lack accountability!

Tender care for world resources IS WITHIN our human forces.
It's our responsibility to have accountability! It's in our hands to help our lands;
BOTH air and water MUST FLOW FREE for ALL EARTH'S WORTH and dignity!"

"When ITS source and course is TETHERED
our treasured, PRESSURED WATER flow...
the WASTE KEEPS ON ITS WRONG, THIS BLOW!
In sum, become more precious!
IF WASTE doesn't haste and IT isn't traced...
if not ceasing, decreasing
but continues INCREASING
taking ITS TOW and IT WILL, we KNOW
as PROBLEMS GROW!"

Her parents told Rosey, "It's all in the KNOWING
it's UP to the people to STOP
DROUGHT FROM GROWING!
People MUST CARE about waters we share:
TO BE CAREFUL WITH USE
'REFUSE AND TO CHOOSE'
VOICING THEIR CHOICE...
NOT TO ABUSE!"

Despite being curious, Rosey got FURIOUS
while listening to them about this MAYHEM!
"To take MORE than you need leads to EXCEED!
To DO SO IS GREED…
THAT'S A PROBLEM INDEED!
To NOT think of others
and dare NOT TO CARE
is a problem for ALL in this
world that WE SHARE."
To Rosey it seemed this
DISCOURSE OF ACTIONS:
Need immediate, expedient,
MUCH NEEDED extractions!
With intention MORE than mention—
VAST AND FAST RETRACTION!
This problem MUST MEND…

This PROBLEM must end!

We NEED to amend INDEED to defend!
So, she took off to visit Roughest her friend.

Rosey shared 'ALL SHE KNEW WITH HIM'
feeling confident together... they'd discuss
and find a STRATAGEM!
Roughest's FREETHINKING
is constantly linking;
he's supportive and causative...
turns NEGATIVE to POSITIVE!
They've been friends for SO LONG
building bonds through the years:
GROWING STRONG - RIGHTING WRONG!
She KNEW Roughest would help her...
THIS PROBLEM would solve!
That working together would
help to RESOLVE;
THE ANSWER TO DROUGHT WOULD
SPROUT, COME ABOUT!
That working with others IS best,
THERE'S NO DOUBT!

Roughest told Rosey, "It's no good to pout
Let's go out and shout to the world about DROUGHT!
How ITS HURTING is GLOBAL. High time to be NOBLE!
We must shout out to others
THIS PROBLEM THAT GROWS...
we MUST tell ALL the world so that

EVERYONE KNOWS!"
QUITE GRIM, she said to him, "This IS a CAUSAL STEAD!
In the past we've sped using our head,
coming up with ideas to end MOST DREAD!

Rectifying, notifying: NOT by implying...
BY OUT SPOKEN advising!
Brainstorming, forming ways to share,
NOT forsaking but making others aware!
ALL THIS WE MUST DARE

SO, PEOPLE WILL CARE!
NO longer indulging instead we're divulging

the TRUTH FORSOOTH, SOON *EPOCH-MAKING!*
S T O P mistaking *THIS OUTBREAKING!*
We NEED to FOCUS ON RECKONING:

THWARDED SHORTAGES BECKONING...
ALL being HURLED throughout our world!"

Roughest knows he and Rosey
have had times STORMY!!
When problems came about, they HAD to go out:
ABSOLUTELY NO DOUBT!
When the current handling was inept
and didn't change but simply kept:
At times- there's JUST a MUST
to RECTIFY, MODIFY,
RUSH to THRUST and
to ADJUST!
(They couldn't ignore it, just had to restore it!)
Said Roughest to her, "We KNOW for sure,
this NEEDS a cure, it CAN'T ENDURE!
TRUST me, you see we MUST adjure:
Making SURE that water's pure!"
She gleefully stressed, "YES, I see WE DO agree.
The object of this project cannot wane!
HOLY COW, we've got to start NOW...
a campaign to gain! IT'S NOT IN VAIN!
Make IT EASY to SEE that DROUGHT CAN be:
STOPPED if we ALL start TO CARE.
For IT'S FEARFUL, if NOT CAREFUL
Our world- WILL BE stripped bare!"

Roughest spouting while shouting:
"We MUST AVOW-GO out NOW! Make people AWARE
this EVERGROWING SCARED AFFAIR
so they can tell others THIS PROBLEM we share."
Roughest cheering when hearing Rosey saying,
"IT'S TRUE! WE NEED help from others;
global sisters and brothers!
All of us; me and you-INDEED we CAN be
Ardent and hardened buccaneers TRUE:
A Readied for Action
SAVE WATER CREW!"

He joined in to say, "Of course, there's a way.
Making sure to be careful with water we use
by thinking of ways **NOT TO *WASTE OR ABUSE!***
When we begin thinking...

PROBLEMS START SINKING
The world WILL be stronger and water lasts longer!"
Rosey joined in the shout, "Let's ALL go out.
Expound what's been found, abound all around
and GO ALL ABOUT!"
Said he to she, "SPOT ON! RIGHT ON!
LET'S DO IT NOW: Go out to tout

and shout of *DROUGHT!*
Collectively thinking TOGETHER gives CLOUT!
Without a doubt- we're certain about!"

"LET'S NOT MOSEY", said Rosey!
Let's ALL join together, help ALL things to grow,
searching, researching SO WE ALL CAN KNOW:
Get IDEAS that free us! SAVE WATERS on earth...

WATER causes
NEW birth!"

How can YOU help OUR waters:::
THE WATERS THAT FLOW?
What CAN YOU and ALL DO:::
ensure that FOR SURE: that
ALL THINGS WELL GROW?
By merging and surging into research,
binding our minds and finding ALL kinds
of ways HAVING WORTH!
Caring and sharing some FOUND 'CURES FOR SURE'
that CAN and WELL help RESTORE, SAVE OUR EARTH
Here's a few samples, a few known examples
needing attention to mention and focus:::
that REALLY should coax us
to join in together, get ideas to use:
IGNITED, EXCITED TO STOP THIS DIFFUSE!

~ ~ ~ ~

Fires! Destroying homes of living things
as animals, plants and human beings!
If there's a fire BUT NOT ENOUGH water to STOP IT,
when *FLAMES GAIN AND BANE,* CAN they be contained?
HOW CAN FIRE CREWS CROP IT? What will these crews do?
IF NOT ENOUGH WATER: TO SUBDUE?

ANY IDEA?

ANY CLUE?

How will homelands and woodlands last or avoid...
if they're marred and charred, scorched or torched,
in sum, become DEVOID BESTROYED?
Fires BLAZE DISMAY– while more and MORE
get BURNT...TORN DOWN with HEATED AMBERS
on the ground: How CAN THESE FIRES be put out
if there's NO WATER TO USE about?

WATERCOURSE RESOURCES
come from different sources-different forces!

SOURCES becoming LIMITED!

SOON MAY BE *PROHIBITED!*

Letting water run and flow

WASTES water used and takes *ITS* toll!

What of '*ELECTRICITY*'
helping everyone to **SEE?**
Plus, cooking, lighting shining bright!
Powering devices we have in our homes,
TVs, appliances, PCs, and phones!

HYDROPOWER! A forceful source SO important,
YES of course! TURBINES GRIND PROMOTING WATER...

~ ELECTRICITY ~ GETS ITS POWER!"

Roughest sighing, replying, "What about the wildlife that live in watered habitats? So many ponds, rivers, seas and streams

it seems: In sum become *POLLUTED VATS!*
They depend on water to live and be!
By polluting their water with treats, you see:
Not knowing better, the treat you give "by their entreat"

will cause resources to *DEPLETE*

and in the end *CAUSE THEIR DEFEAT!*
Our friends with wings: Ducks, pigeons, geese:
Get broken bodies! *MOBILITY CEASED!*
Wings and feet may be *DEFORMED...*
Our good intentions **only** *HARMED!*

DAMAGES done *HINDERS* migration:

NATURE'S *IMPAIRED*, a *SAD SITUATION!*"

"If they can't leave- NOT able for stable
and NATURAL RELOCATION;
They soon create a state of HARMFUL HABITATION:
EVADED; DEVASTATED-SPILLED OVER POPulation!
DAMAGED AND FRAYED,
POLLUTED -DECAYED!"
PROBLEM SOLVING, INVOLVING
(Oh, so true, ME and YOU!)
POLLUTED WATERS EFFECT LAND, TOO!
We really CANNOT hesitate!
Stand by and sigh...JUST HOPE and WAIT!
Animals and humans throughout the world,
plants and trees all GET DISEASED!
WATER WASTE ATTACKS are HURLED...
WE'VE GOT TO DEFEAT
NUTRIENT POLLUTION
WE'VE GOT TO, OUGHT TO
FIND SOLUTION!"

Rosey said with resolution:
"It's HIGH TIME for RESTITUTION!
Other THINGS that people bring
WRING HARM, ALARM and ZING WITH STING:
When cleaning hands and teeth OR washing dishes
do YOUR part…SHUT faucets OFF,
UNTIL it's time for RINSES-SWISHES!
Just letting it flow: You know YOU SHOULDN'T ought to-
could be NEXT TIME you'll find YOU'RE OUT of water!
Fears of global warming swarming: Making it hotter:
Discussions gush of repercussions: Prompt confusion-
CONVOLUTION in this 'climate teeter-totter'…

ITS reported and ITS sighted; recorded and indicted:
ITS POLLUTING GLOBAL WATER!"

Yes, the facts ARE real
IT'S A 'WHEELED' BIG deal!

Folks it's UP to US to bring *UNITY*

and **DO SOMETHING!**
Get together, one another

unite and fight this *GROWING PLIGHT*
as global sisters, brothers.
Do it NOW, for WE CAN'T WAIT
NO LONGER CAN WE HESTITATE!
THIS EVER-GROWING HARMFUL STATE:
CONTAMINATION, DEPRIVATION...
THIS ONGOING DEVASTATION!
LET'S PROBE TO KNOW TO HELP OUR GLOBE:
You KNOW we OUGHT TO...
SAVE WORLD

WATER!

My! We've read and heard alarming words to help us think, to help us link:
Articles, though similar- prompt us to be familiar!
They help to reach, beseech to teach with facts and acts that
have impact; directing where and how to act.

There's SO much MORE we should explore to change WORLD SITUATION.
Instead of shirking, let's start working TOGETHER save creation!
Yes, ALL of US can set the stage, help save our world, just turn the page!

Readers: We've come to the end
of this water flow story.
Yet, sure NOT the end that
threatens world glory!
If you respond
and want to move on
you'll find photos and charts
that get to the heart
of what's needed to know

to stop *THIS HARM* grow!
Use of rhyming that's staggered
with rhyming in time
will help to promote, so you CAN
devote YOUR part to end crime!

THIS crime against humanity calls for HUMAN SOLARDARITY: Photos, articles, some particles of humor, show what's happening with polluted-wasted water so maddening and saddening. Unite in the fight of this plight that's in sight. Gather with others to make all things right!

Teachers in the classroom 
Teachers in the home 
Learners and leaders
throughout the global roam.
Scientific history here to see for you and me,
All these facts have value-worth
they can help us save our earth!
Worth the read? Yes, indeed!
We can help our world succeed!
Life on earth indeed does need
our pledge and promise...keep our creed.
We must take the time to end this crime
STOP this growing genesis

SWARMING, HARMING MENACES
These found actual, factual premises
we need indeed send messages.
So take your time and preferences
to read and hear these references.

Let's take a closer L👀K...
from what we've learned within this book.
Lurch into research! Indeed, need to read!
L👣K to *SEE* what's been found...
In the sky, reasons why:
ON LAND; IN WATER; UNDERGROUND:
What we should know about waters that flow!

IT'S FACTUAL, OUR NATURAL water cycle HELPS clean the air and nourish land;
Polluting mayhem does condemn; this problem grows~ IT'S OUT OF HAND!
ALL this upsets the ~RHYTHM within LIFE'S ECOSYSTEM~
Our ecosystem DEPENDS ON ITS FLOW to HELP the sun SO ALL WILL GROW!

FROM YOUNG TO OLD IT MUST BE TOLD
SO WE CAN KNOW; HELP WATERS GROW AND HOW TO SHOW.
SO 'VITAL FOR ALL to RESOLVE'...THIS WATER SITUATION!
Pollution Solution CAN SOLVE through EDUCATION

HYDROPOWER (from Greek: ὕδωρ, "water"), also known as water power, is the use of falling or fast-running water to produce electricity or to power machines. This is achieved by converting the kinetic energy of water into electrical or mechanical energy. Hydropower is a form of sustainable energy production.

HOW DO YOU MAKE ELECTRICITY FROM WATER?

Hydroelectric power stations work when flowing water runs through a turbine, which spins the rotor of an electricity generator and creates a magnetic field that induces an electric current. A turbine spins a rotor to produce electricity from water.

HOW CAN WATER BE USED TO CREATE ELECTIRICITY?

Electricity generated from water on the ocean is known as wave power or wave energy. This method of power generation uses changes in the air levels of sealed chambers to power turbines. These chambers are floated on parts of the ocean with high wave activity, ensuring that a great deal of electric energy can be produced.

When a dam is used to generate energy, tunnels are installed in the dam when it is built. These tunnels are lined with turbines which are turned when water flows through the tunnels. As the turbines turn, they create electricity which can be fed into the grid or stored. Dam operators can determine the amount of energy produced by regulating the flow of water; most dams are capable of generating far more power than they do on a daily basis, which can be useful when there are problems at other power plants and facilities.

Electricity generated from water on the ocean is known as wave power or wave energy. This method of power generation uses changes in the air levels of sealed chambers to power turbines. These chambers are floated on parts of the ocean with high wave activity, ensuring that a great deal of electric energy can be produced. Not all areas of the ocean are suitable for the generation of wave power, but some seaside communities have taken advantage technology to power themselves.

Electricity generation is a major concern for much of the world, since demand is only rising with the growing human population. The benefit of hydroelectric power is that once generation facilities are built, it is easy to maintain and operate them. Electricity generated from water is also clean, since it doesn't involve the burning of fossil fuels to generate power. People can also generate hydroelectric power themselves, if they have access to a fast-moving body of water so that they can install waterwheels.

Using the energy potential of trapped water in a dam is one way to generate electricity from water.

HOW IS WATER POWER HARNESSED TO PRODUCE ELETRICTIY?

To harness energy from flowing water, the water must be controlled. A large reservoir is created, usually by damming a river to create an artificial lake, or reservoir. Water is channeled through tunnels in the dam. The energy of water flowing through the dam's tunnels causes turbines to turn. The turbines make generators move.

Electricity generated on the ocean is known as "wave power"

There are some drawbacks to electricity generated from water. Dams, for example, can be quite destructive when they are installed, as water will flood the regions behind dams. This has been a cause for controversy in the past, especially when dams flood valleys used by native peoples for burial and religious ceremonies. If a dam fails, it also causes catastrophic flooding and people downstream of a dam tend to experience a reduction in available water after it has been installed. Concerns have also been raised about wave power, since it can be quite noisy and it may prove damaging to marine life.

WATER POLLUTION FACTS:

80% of pollution in seas and oceans comes from land.
Some amount of water pollution is due to air pollution. The pollutants from air settle on water thus polluting it.
Small drops of motor oil that drip from automobiles on roads eventually find their way to the sea.
A third of water with shellfish in the US is contaminated with pollutants.
A gallon of paint can pollute 250,000 gallons of water while a gallon of gasoline holds the potential to pollute 750,000 gallons of water.

WHAT IS WATER POLLUTION?

Water pollution is defined as any change in the water that renders it unusable and harmful for living organisms. In other words, water gets so bad that you cannot drink it, bathe in it, wash clothes, or give it to animals.

Water gets polluted when unwanted substances enter it and change its composition. The unwanted material is what we call a pollutant. Not just freshwater but seawater, too, is prone to several types of pollution.

TYPES OF WATER POLLUTION

Water pollution can be broadly classified into two types – point source, where the source is identifiable and non-point source pollution, where the contaminant seems to be coming from throughout the place across the landscape. Water pollution can be elaborately classified depending on the cause of water pollution.

Organic waste: This occurs due to the dumping of organic matter in the water, such as draining sewage water into the water bodies. The organic matter can bring down the oxygen level in the water thus making it unfit for underwater organisms

Chemical: Examples include effluent water coming from pesticides factory and industries that process heavy metals. Another form of chemical water pollution is oil leakage that causes an oil layer to persist on top of the water. Chemical pollutants are toxic to all life forms.

Microbiological: Water could contain potential pathogen like a virus, which causes epidemics. For example, untreated sewage water coming out from an area might contain an epidemic; similarly, untreated water coming out from laboratories that conduct biological experiments and follow no safety norms can be unsuitable for use.

Acidity: Acidic pollution happens when a contaminant changes the pH of the water. Organisms that live in water are sensitive to pH levels. Any change to the pH level due to the addition of a substance is water pollution.

Nutrient: Excessive fertilizers from agricultural land can get into water bodies making the water abnormally rich in nutrients. It can lead to a surge in the population of algae and other organisms, which can affect the population of other living things in the water body.

Sedimentation: Eroded sediment, such as in the areas of mines, can get into the water and change its quality.

Thermal: Thermal water pollution happens when human activities lead to an abnormal change in the temperature of water. For instance, a factory may dump hot water into a nearby stream thus increasing its temperature. Living organisms are sensitive to temperatures.

It is clear that human activities are the leading cause of water pollution. Human beings render water unusable not just for themselves but also for other living organisms.

WHAT ARE THE EFFECTS OF WATER POLLUTION

The impact of water pollution is felt at multiple levels: Here are the most common effects:

Impact on wildlife: Plastic waste dumped into water bodies kill over a million birds and animals in a year. Plastic is the most common pollutant that children are likely to know. There are several other pollutants as well. For instance, the road salt used for melting accumulated ice on roads contains an anti-caking agent called ferrocyanide, which can seep into the groundwater and lakes through melted ice. Cyanide is poisonous to most life forms, and the presence of ferrocyanide can destroy water flora and fauna.

Change in ecosystems: The long-term impact of water pollution is the widespread change in the ecosystem. For instance, water pollution can change the entire habitat of a region due to the rise of one type of population. A change in the ecosystem can affect human beings too.

Spread of diseases: Environmental experts state that waterborne illnesses account for 80% of all infectious diseases in the world. About 80% of sewage in developing countries is released untreated into water bodies. Consumption of water polluted with organic waste is a common cause for illnesses. Children are the most vulnerable to such infections.

A decline in human health: Humans pollute the water and drink it too. Drinking polluted water can cause a host of diseases and chronic illnesses. For instance, nitrate is a common chemical contaminant in groundwater. Children who drink water high in nitrates can suffer from a variety of problems. Consuming water rich in algae can cause liver, neurological and respiratory problems. Heavy metals from polluted water can stay within the body for a long time and create a host of problems.

Mounting economic costs: Did you know? The dirtier the water, the more money your local government needs to spend to treat it and make it potable. The tourism industry in the US loses $1 billion each year due to water pollution because the water is unsuitable for activities like boating. The fishing industry is negatively impacted too.

Properties close to the polluted lakes suffer a severe drop in value thus affecting the real estate business. All this can have collateral effects like loss of jobs and opportunities for people employed in such sectors.

Water pollution is a serious issue. But you can make it simpler for the kids to understand through some activities.

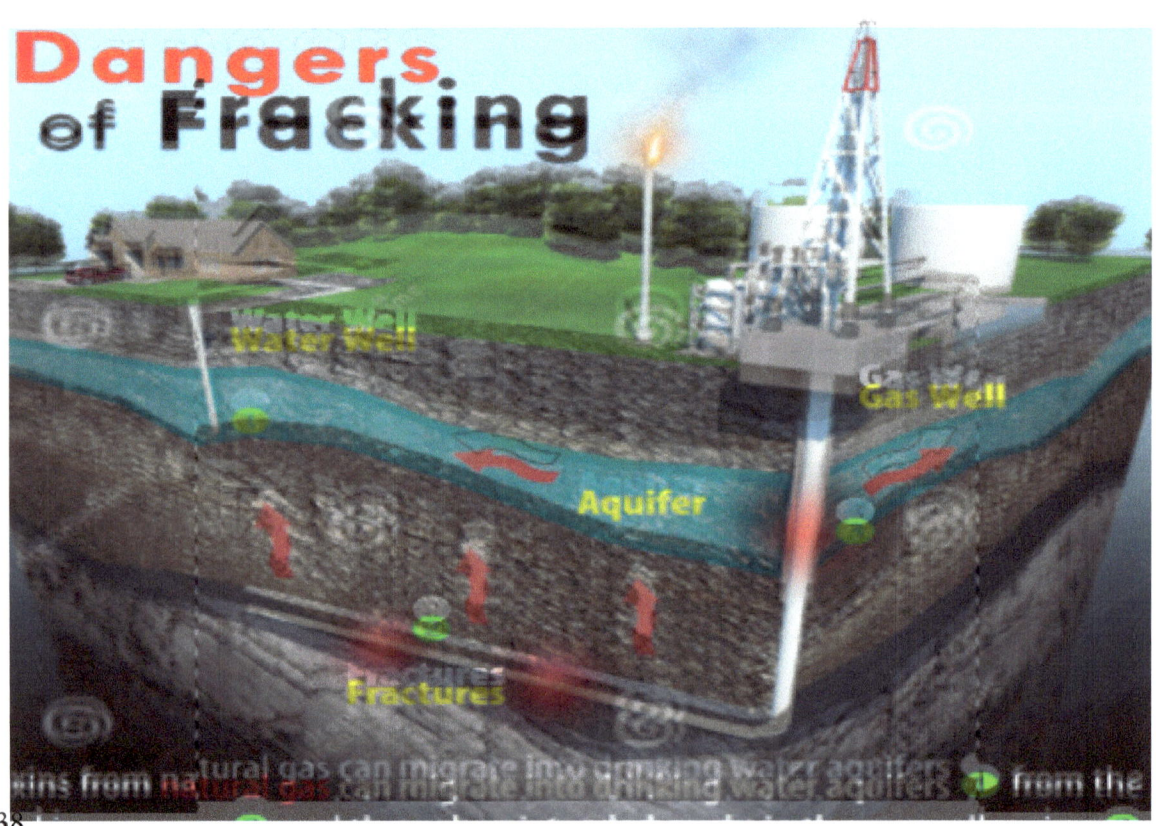

References Article 1: Pgs 56-62

How to Prevent Water Pollution?

Preventing water pollution is easier than it seems. Children can adopt simple, everyday best practices. Here is what you can do to prevent water pollution:

Never dump garbage in water bodies: The next time the child goes for a boat ride, let them not throw thrash into the water body. Organic matter like flowers or fruit skin is a water pollutant! Teach children to carry a paper bag along and use it to dispose of the thrash.

Do not openly defecate or urinate near water: If you are at the lakeside for a picnic, then tell the child that it is not right to answer nature's call anywhere around the lake. Direct excretion of organic waste even around the lake can cause the waste to find its way into the water and pollute it.

Dispose of liquid wastes carefully: If you have liquid chemicals, oil-based chemicals, or any other liquid compound that could be a potential pollutant, then do not throw them into the drain right away. Your local sewage treatment plant may not be able to eliminate harmful chemicals before dumping it into a water body. Instead, consult a local government department or an organization that can dispose of such compounds in an environmentally friendly way.

Use eco-friendly pesticides and fertilizers: For instance, you can use physical barriers like a mesh to protect plants from pests instead of using chemical pesticides. Several plant-based, natural fertilizers provide only as many nutrients as plants need. It prevents nutrients drain off to water bodies where it can cause pollution.

39

Teach environment-friendly habits: You can reuse the water left after washing grains and vegetables, for watering plants. If you have a garden, then use the water to make your own compost through all the vegetable and plant waste generated in your home. The compost is an excellent eco-friendly fertilizer. Good practices encourage children to minimize the amount of waste that potentially finds its way into a water body.

Water is what *sustains the life for the tiniest to the largest living organism on the planet. Water pollution affects every creature, and we human beings are single-handedly responsible for water contamination*

WHAT IS A POLLLUTANT?

A pollutant is a substance or energy introduced into the environment that has undesired effects, or adversely affects the usefulness of a resource. A pollutant may cause long- or short-term damage by changing the growth rate of plant or animal species, or by interfering with human amenities, comfort, health, or property values. Some pollutants are biodegradable and therefore will not persist in the environment in the long term. However, the degradation products of some pollutants are themselves polluting such as the products DDE and DDD produced from the degradation of DDT.

5 Ways Water Pollution Is Killing Animals

(This article was first published by Care2.com on 20 Aug 2017. September 3, September 3, 2017 Supertrooper News, Wildlife)

Human activities are contaminating the world's water systems and disrupting wildlife. From toxic chemical runoff to the accumulation of litter miles away from land, here are five ways water pollution is killing animals:

1. More Than a Million Sea Birds and Mammals Are Killed Each Year by Ingestion of Plastic Around 100 million tons of plastic i**s produced every year, of which 10 percent ends up in the sea. Ocean litter comes from many sources — including trash that washes off city streets, waste blown in from landfills and containers that fall off ships during heavy storms. Once in the water, the debris accumulates in large patches, travels with currents and washes up onshore. This litter is frequently consumed — often with fatal effects — by marine mammals, fish and birds who mistake it for food. The Trash Vortex of the North Pacific Ocean consists of trillions of pieces of decomposing plastic that forms a gigantic swirling garbage patch the size of Texas.

2. Chemical Runoff from Farms Has Caused 400 Dead Zones Around the World

The use of pesticides and fertilizers on farms has increased an alarming 26-fold over the last 50 years, causing serious environmental consequences. Chemical runoff from farms is leaching into nearby streams, waterways and groundwater, killing thousands of insects and fish. The presence of fertilizers in the water alters nutrient systems, resulting in explosive growths of algae that produce harmful toxins and deplete the water of oxygen, and as a result little or no marine life can exist. Scientists have recorded an estimated 400 such dead zones around the world.

3. Noise Pollution Is Driving Animals to Extinction

Pollution is not always physical. Sound waves from ships, sonar devices and oil rigs can travel for miles disrupting migration, communication, hunting and reproduction patterns of many marine animals. The deafening noise of gas and oil explorations are so loud that they are causing devastating effects to the sea life residing in our world's oceans such as mass strandings, reckless diving, the inability to find food and chronic stress. Extreme noise pollution has been known to kill hundreds of dolphins and whales at a time, many of which are already on the brink of extinction.

4. Cruise Ships Dump More Than 250,000 Gallons of Wastewater and Sewage Every Day

Due to lax laws, cruise ships have been operating with little to no environmental regulations, and as a result have caused a great deal of damage to sensitive marine life. Current regulations allow cruise ships to legally dump untreated sewage and other waste once the ships are three miles from shore. This toxic waste is discharged directly into the ocean and contains bacteria, pathogens, medical waste, oils, detergents, heavy metals and other harmful substances, all of which are putting aquatic life at risk.

5. Acid Rain Discharges Toxic Amounts of Aluminum into the Water Systems

When water in the atmosphere mixes with certain chemicals, in particularly those omitted from burning fossil fuels, acidic compounds are formed. Acid rain has been linked to many serious adverse effects on ecosystems, especially aquatic ecosystems on which it falls. Through the discharge of toxic amounts of aluminum into the water, pH levels are altered, killing many animals outright and throwing delicate ecosystems out of balance.

You can help to protect the planet and sustain the world's ecosystems by supporting environmental groups that are fighting to put a stop to these harmful practices, as well as by making your own conscious decisions regarding waste management, harmful chemicals and earth friendly alternatives to your everyday products earth friendly alternatives to your everyday products.

This article was first published by Care2.com on 20 Aug 2017.

The Negative Effects of Water Pollution on Animals

Considering the negative effects that humans endure, because of water pollution, there is no doubt that the same there are similar effects left on animals, both on land and residing in the water.

The reality of water pollution and the effect it has on animals is quite saddening. The thought that water is only polluted by the littering of plastic and man-materials and chemicals, without considering one of water pollution's main sources, is also quite scary. Air pollution, smog, and chemicals that ponder in the air, because of industrialized activity, causes more harm to the ocean and useful water sources than people might think.

While humans only feel the consequences of water pollution when it affects the water they consume, animals that, apart from drinking water, make it their habitat, is at extreme risk of illness, as well as dying as a result thereof. While humans have knowledge and awareness about polluted water and that they should avoid it at all costs, animals do not have that luxury and are forced to endure the effects of pollution every single day.

4 Ways polluted water effects animals

Water pollution causes an excess of minerals and nutrients, like phosphorous and nitrogen in the water, that leads to increased growth of aquatic plants and algae that can be quite toxic, cause poisoning and whichever fish or animals that feed on it, to die.

Polluted water may contain chemical contaminants, that are found in industrial wastes, that can kill aquatic organisms, such as fish, frogs, and tadpoles. It also causes a major loss of food for bigger aquatic organisms.

Pollution that is caused by oil spills exposes unhealthy levels of oil to marine life, that results in the marine environment to become toxic, eventually causing marine animals to die.

Polluted water, used for irrigation, contaminated the soil, as well as agricultural produce, which furthermore leads to health issues in cattle and herbivorous animals.

Water pollution and human health

Mehtab Haseena*, Muhammad Faheem Malik, Asma Javed, Sidra Arshad, Nayab Asif, Sharon Zulfiqar and Jaweria Hanif

Department of Zoology, University of Gujrat, Pakistan

Corresponding Author: Mehtab Haseena, Department of Zoology, University of Gujrat, Pakistan

Accepted Date: July 13, 2017 DOI: 10.4066/2529-8046.100020

Visit for more related articles at Environmental Risk Assessment and Remediation

Abstract

This study was conducted at Department of Zoology, University of Gujrat, Pakistan during 2016-2017 as a term paper for Master of Philosophy. The data regarding water pollution and human health was obtained and compiled through a thorough review of various published research articles of international reputed journal and relevant books. Water covers about 70% Earth's surface. Safe drinking water is a basic need for all humans. The WHO reports that 80% diseases are waterborne. Industrialization, discharge of domestic waste, radioactive waste, population growth, excessive use of pesticides, fertilizers and leakage from water tanks are major sources of water pollution. These wastes have negative effects on human health. Different chemicals have different affects depending on their locations and kinds. Bacterial, viral and parasitic diseases like typhoid, cholera, encephalitis, poliomyelitis, hepatitis, skin infection and gastrointestinal are spreading through polluted water. It is recommended to examine the water quality on regular basis to avoid its destructive effects on human health. Domestic and agriculture waste should not be disposed of without treating.

Keywords: Water pollution, sources of water pollution, harmful chemicals, infectious diseases

Introduction

Water pollution occurs when unwanted materials enter in to water, changes the quality of water [1] and harmful to environment and human health [2]. Water is an important natural resource used for drinking and other developmental purposes in our lives [3]. Safe drinking water is necessary for human health all over the world. Being a universal solvent, water is a major source of infection. According to world health organization (WHO) 80% diseases are water borne. Drinking water in various countries does not meet WHO standards [4]. 3.1% deaths occur due to the unhygienic and poor quality of water [5].

Discharge of domestic and industrial effluent wastes, leakage from water tanks, marine dumping, radioactive waste and atmospheric deposition are major causes of water pollution. Heavy metals disposed of and industrial waste can accumulate in lakes and river, proving harmful to humans and animals. Toxins in industrial waste are the major cause of immune suppression, reproductive failure and acute poisoning. Infectious diseases, like cholera, typhoid fever [6] and other diseases gastroenteritis, diarrhea, vomiting, skin and kidney problem are spreading through polluted water [7]. Human health is affected by the direct damage of plants and animal nutrition. Water pollutants are killing sea weeds, mollusks, marine birds, fishes, crustaceans and other sea organisms that serve as food for human. Insecticides like DDT concentration is increasing along the food chain. These insecticides are harmful for humans [8].

Major sources of water pollution:
1) Domestic Sewage 2) Industrialization 3) Population growth 3) Pesticides and fertilizers
4) Plastics and polythene bags 5) Urbanization and 6) Weak management system

It is reported that 75 to 80% water pollution is caused by the domestic sewage. Waste from the industries like, sugar, textile, electroplating, pesticides, pulp and paper are polluting the water [9]. Polluted rivers have intolerable smell and contains less flora and fauna. 80% of the world's population is facing threats to water security [8].

Large amount of domestic sewage is drained in to river and most of the sewage is untreated. Domestic sewage contains toxicants, solid waste, plastic litters and bacterial contaminants and these toxic materials causes water pollution. Different industrial effluent that is drained in to river without treatment is the major cause of water pollution [9]. Hazardous material discharged from the industries is responsible for surface water and ground water contamination. Contaminant depends upon the nature of industries. Toxic metals enter in to water and reduced the quality of water [10]. 25% pollution is caused by the industries and is more harmful [11].

Increasing population is creating many issues but it also plays negative role in polluting the water [10]. Increasing population leads to increase in solid waste generation [12]. Solid and liquid waste is discharged in to rivers. Water is also contaminated by human excreta. In contaminated water, a large number of bacteria are also found which is harmful for human health [11].

Government is incapable to supply essential needs to citizens because of increasing number of populations. Sanitation facilities are more in urban areas than rural areas. Polythene bag and plastic waste is a major source of pollution. Waste is thrown away by putting it in to plastic bags [11]. It is estimated that three core people of urban areas defecate in open. 77% people are using flush latrines and 8% are using pit latrines. Urbanization can cause many infectious diseases. Overcrowding, unhygienic conditions, unsafe drinking water are major health issues in urban areas. One quarter of urban population is susceptible to disease [9]. Pesticides are used to kill bacteria, pest and different germs. Chemical containing pesticides are directly polluting the water and affect the quality of water. If pesticides are excess in amount or poorly managed then it would be hazardous for agriculture ecosystem [13,14]. Only 60% fertilizers are used in the soil other chemicals leached in to soils polluting the water, cyanobacteria are rich in polluted water and excess phosphate run off leads to eutrophication.

Residues of chemicals mix with river water due to flooding, heavy rainfall, excess irrigation and enter in the food chain. These chemicals are lethal for living organisms and many vegetables and fruits are contaminated with these chemicals [9,15]. Trace amounts of pharmaceutical in water also causes water pollution and it is dangerous to human health [16].

Effects of water pollution on human health

There is a greater association between pollution and health problem. Disease causing microorganisms are known as pathogens and these pathogens are spreading disease directly among humans. Some pathogens are worldwide some are found in well-defined area [9]. Many water-borne diseases are spreading man to man [17]. Heavy rainfall and floods are related to extreme weather and creating different diseases for developed and developing countries [18]. 10% of the population depends on food and vegetables that are grown in contaminated water [19].

Many waterborne infectious diseases are linked with fecal pollution of water sources and results in fecal-oral route of infection [20]. Health risk associated with polluted water includes different diseases such as respiratory disease, cancer, diarrheal disease, neurological disorder and cardiovascular disease [21].

Nitrogenous chemicals are responsible for cancer and blue baby syndrome [22]. Mortality rate due to cancer is higher in rural areas than urban areas because urban inhabitants use treated water for drinking while rural people don't have facility of treated water and use unprocessed water.

Poor people are at greater risk of disease due to improper sanitation, hygiene and water supply [12]. Contaminated water has large negative effects in those women who are exposed to chemicals during pregnancy; it leads to the increased rate of low birth weight as a result fetal health is affected [23].

Poor quality water destroys the crop production and infects our food which is hazardous for aquatic life and human life [7]. Pollutants disturb the food chain [17] and heavy metals, especially iron affects the respiratory system of fishes. An iron clog in to fish gills and it is lethal to fishes, when these fishes are eaten by human leads to the major health issue [24]. Metal contaminated water leads to hair loss, liver cirrhosis, renal failure [25] and neural disorder [26].

Bacterial diseases

Untreated drinking water and fecal contamination of water is the major cause of diarrhea. *Campylobacter jejuni* spread diarrhea 4% to 15% worldwide. Fever, abdominal pain, nausea, headache are major symptoms of diarrhea. Good hygienic practices and use of antibiotics can prevent this disease.

Disease cholera is caused by the contaminated water. *Vibrio Cholerae* is responsible for this disease. This bacterium produces toxins in digestive tracts. The symptoms of this disease are watery diarrhea, nausea, vomiting and watery diarrhea leads to dehydration and renal failure. Anti-microbial treatment is used to get rid of this disease.

Shigellosis is a bacterial disease caused by *Shigella bacteria*. It affects the digestive tract of humans and damages the intestinal lining. Watery or bloody diarrhea, abdominal cramps, vomiting and nausea are symptoms and it can be cured with antibiotics and good hygienic practice. Salmonellosis infects the intestinal tract. *Salmonella* bacteria are found in contaminated water and it results in inflammation of intestine and often death occurs. Antibiotics are prescribed for this disease [27].

SHAPES OF BACTERIA

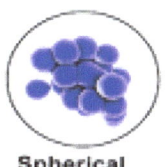 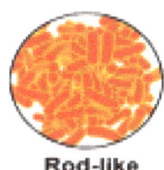 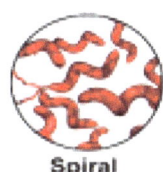

Spherical   Rod-like   Spiral

Viral diseases

Hepatitis is a viral disease caused by contaminated water and infects the liver. Jaundice, loss of appetite, fatigue, discomfort and high fever are symptoms of hepatitis. If it persists for a long time, it may be fatal and results in death. Vaccine is available for hepatitis and by adopting good hygienic practice, one can get rid of this disease [27].

Encephalitis is inflammatory disease spread by bite of infected mosquitoes. Culex mosquito lays their eggs in contaminated water. Most people don't show any symptoms but some symptoms are headache, high fever, muscle stiffness, convulsions however in severe cases coma and paralysis results. No vaccine is available for this disease [28].

Poliomyelitis virus is responsible for poliomyelitis. Sore throat, fever, nausea, constipation and diarrhea and sometimes paralysis are symptoms of poliomyelitis. Vaccine is available for this disease [28]. Gastroenteritis is caused by different viruses including rotaviruses, adenoviruses, caliciviruses and Norwalk virus. Symptoms of gastroenteritis are vomiting, headache and fever. Symptoms appear 1 to 2 days after infecting. Sickness can be dangerous among infants, young children and disabled person [28].

Parasitic diseases

Cryptosporidiosis is a parasitic disease caused by the *cryptosporidium parvum*. It is worldwide disease and symptoms are diarrhea, loose or watery bowls, stomach cramps and upset stomach [28]. *Cryptosporidium* is resistant to disinfection and affects immune system and it is the cause of diarrhea and vomiting in humans [29].

Galloping amoeba is caused by the *Entamoeba histolytica* and affects stomach lining. This parasite undergoes cyst and non-cyst form. Infection occurs when cyst found in contaminated water and it is swallowed. Symptoms are fever, chills and watery diarrhea [27]. According to WHO, diarrheal cases are about 4 billion and results in 2.2 million deaths [30].

Giardiasis is caused by *Giardia lamblia*. Cells of intestinal lining may become injure. *Giardia* is resistant to wintry temperature and disinfectant. Sometimes it is known as travelers' disease. People suffering from giardiasis have symptoms bloating, excess gas, watery diarrhea and weight loss [28].

Conclusion and recommendations

Water pollution is a global issue and world community is facing worst results of polluted water. Major sources of water pollution are discharge of domestic and agriculture wastes, population growth, excessive use of pesticides and fertilizers and urbanization. Bacterial, viral and parasitic diseases are spreading through polluted water and affecting human health. It is recommended that there should be proper waste disposal system and waste should be treated before entering in to river. Educational and awareness programs should be organized to control the pollution.

References Article 2: Pgs.55-61

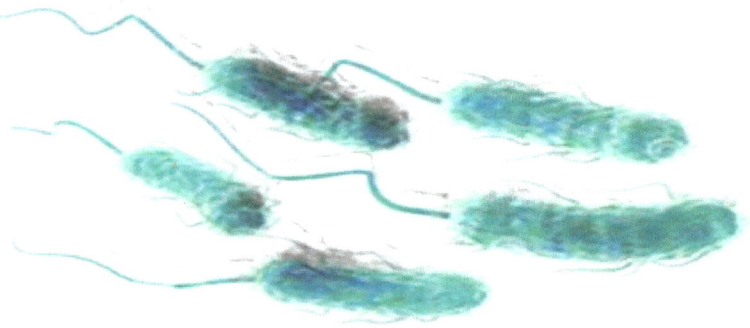

Water pollution is indeed a massive problem all around the world. It's a known fact that our planet is 70 percent water. For this reason, water pollution is certainly a pain in the ass. It affects not just the rich marine life, but more importantly, us, humans. Battling it has always been a forefront endeavor of nation leaders ever since. Countless laws and acts as well as rallies have always been made to ensure that this problem isn't just being taken for granted. Countries have definitely been talking and dealing about it together as one. Unfortunately, it's a challenge that seems to stand the test of time. Today in this blog post, we're going to talk about the effects of water pollution on aquatic animals. Make sure to stay tuned until the very end.

Contents

What is Water Pollution?

Various Types of Water Pollutants

1. Organic Materials

2. Plant Nutrients

3. Toxic Pollutants

4. Physical Pollutants

5. Biological Pollutants

Effects of Water Pollution on Aquatic Animals

Final Thoughts

What is Water Pollution?

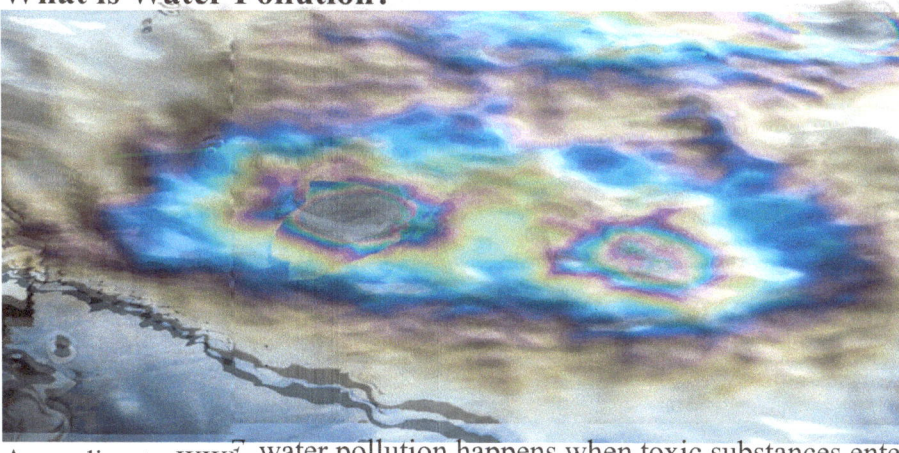

Airiel view

According to *WWF*, water pollution happens when toxic substances enter water bodies such as lakes, rivers, oceans, and so on. What actually happens is that these substances dissolves in the water. After that, two things may happen: they can either stay suspended in the water or gets deposited in the bed or water floor. In general, these pollutants degrade the quality of the water. The most unfortunate thing that can indeed happen is that those pollutants reaching the groundwater. When they do, they'll absolutely reach our homes. Worse of all, it can be in the water that we're drinking. Yes, what an enormous nightmare!

Various Types of Water Pollutants

Generally, the known water pollutants as of now are discharged trash and sewage sludge in courtesy of humans. Of course, there are also the infamous chemical, industrial, and radioactive wastes. Furthermore, **The Open University** has a comprehensive list of water pollutants:

1. Organic Materials

These substances are the majority of the freshwater pollutants. These organic materials come from domestic sewage discharges. Additionally, these can also come from food processing industries. Moreover, take note that this type of water pollutant can be further categorized based upon its form.

With regards to this, there are biodegradable as well as non-biodegradable toxic organic pollutants. Furthermore, these water pollutants can also be natural or synthetic. In the case of natural toxic organic pollutants, they are certainly biodegradable. However, synthetic toxic organic pollutants can be degraded by microorganisms.

Additionally, know that the major polluting effect of these materials is that they reduce the oxygen in the water. What actually happens is that organisms break these down into simpler or inorganic substances. Indeed, along that process, they utilize oxygen. As these organisms increase in number, the need for oxygen increases as well. Certainly, the oxygen concentration in that body of water continues to go down. That right there is a vital problem for the marine animals there.

2. Plant Nutrients

Eutrophication is the increase with time of plant nutrients and biota in a body of water. Certainly, there are inorganic substances that plants need for their metabolism. On the other hand, they can reach a point where they're just too much. That peak level may already imply that they have become pollutants already.

Moreover, do take note that the major polluting effect of these nutrients-turned-to-pollutants is that they accelerate the natural process making algae grow so fast. When the latter happens, algal blooms have now formed and that's a big problem. These algal blooms or the distinct rapid increase of algae in the body of water impose threat to the marine life there. These blooms lead to excess respiration in water. Excess respiration is the dynamic where there's no replacement of oxygen. Thus, the life of many aquatic animals in the ocean is at stake.

Furthermore, the common organisms associated with these algal blooms are the Cyanobacteria and unicellular green algae. The Cyanobacteria is also the blue-green algae. They can fix atmospheric nitrogen. At the same time, they're also not dependent on dissolved carbon dioxide. It's because they can utilize the present bicarbonate ions. On the other hand, unicellular green algae need nitrate. Unlike the Cyanobacteria, they're not able to fix atmosphere nitrogen. In addition to this, they also demand high levels of carbon dioxide.

3. Toxic Pollutants

Toxic pollutants very in their polluting effect towards the body of water involved. First, they can be accumulative which means that their effect increases as doses of them are added more. Second, they can also be chronic in which their effect can be observed within a prolonged period of time. Additionally, this effect can either be lethal or sublethal. Third, toxic pollutants can also be acute. If that's the case, their effect is prominent in just a short period of time. Also, that effect can possibly be death.

Moreover, they can also be sublethal wherein it's not really automatically resulting to death. The most possible cases are indeed a change in behavior, growth, and reproductive success of marine life. Lastly, they can be lethal which is sudden death by direct intoxication. Take note that chemicals in specific dosages can be classified into various toxicity levels.

Furthermore, these chemicals that have the possibility to become poisonous are arsenic, cadmium, copper, lead, and mercury. There's also the aluminum, molybdenum, and zinc. At the same time, inorganic substances such as cyanide, fluoride, sulfide, sulfite, and nitrate can be toxic too.

4. Physical Pollutants

Physical pollutants are suspended solids and immiscible liquids. There are also discharges leading to temperature change or changes in the flow rate of the receiving water. At the same time, there are also substances that give the water a certain color, odor, as well as taste.

Suspended solids or particulates come from many industries. Without a doubt, they reduce the light penetration into the water. Moreover, they sill oppress aquatic animals and plants living on the bottom as they settle. They can indeed disrupt the life underwater. On the other hand, there are also the immiscible liquids. These come from oils as well as greases or tarry substances. It's known fact that oil is less dense than water. For this reason, it'll very likely pollute a very enormous area even in just a small quantity. Another known fact that's extremely unfortunate is that oil is one of the most serious water pollution problems.

Furthermore, the discharges causing temperature change in the water are industrial effluents. These are also the culprit behind the changes in the flow rate of the body of water. Lastly, chemical compounds are the pollutants causing color, odor, and taste changes in the water.

5. Biological Pollutants

These are organisms that when ingested by other forms of marine life can be detrimental. There are the pathogenic bacteria, coliforms, and faecal streptococci. There's also the *Clostridium perfringens*, viruses, protozoa, helminths, and many more organisms.

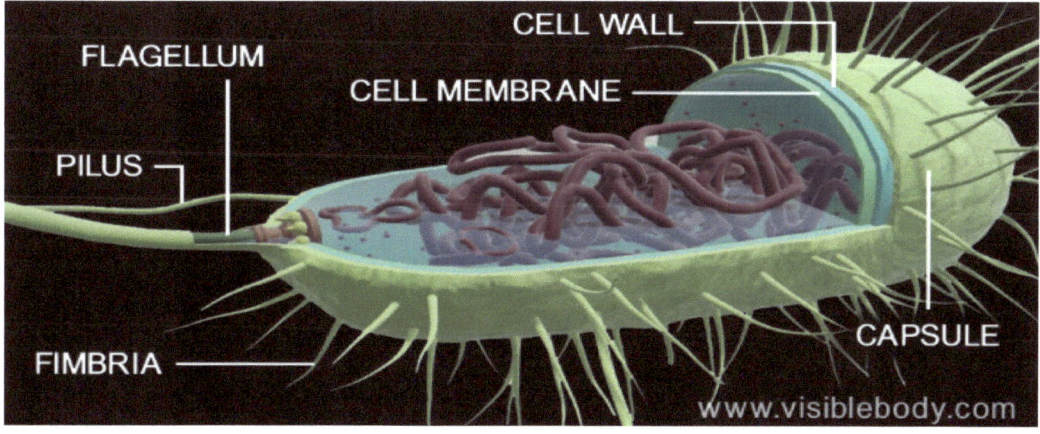

The more we explore, the more we're for sure
that ALL things on earth, have value, self-worth.
All living things depend on what's common of:
living and giving, a natural phenomenon!

Let's take a closer look: What's different? The same?
Animal, plant and human cells can AND do share like names!
Indeed, these all strive and NEED water to thrive
to restore, reproduce, resupply= STAY ALIVE!

Common cells in Plants

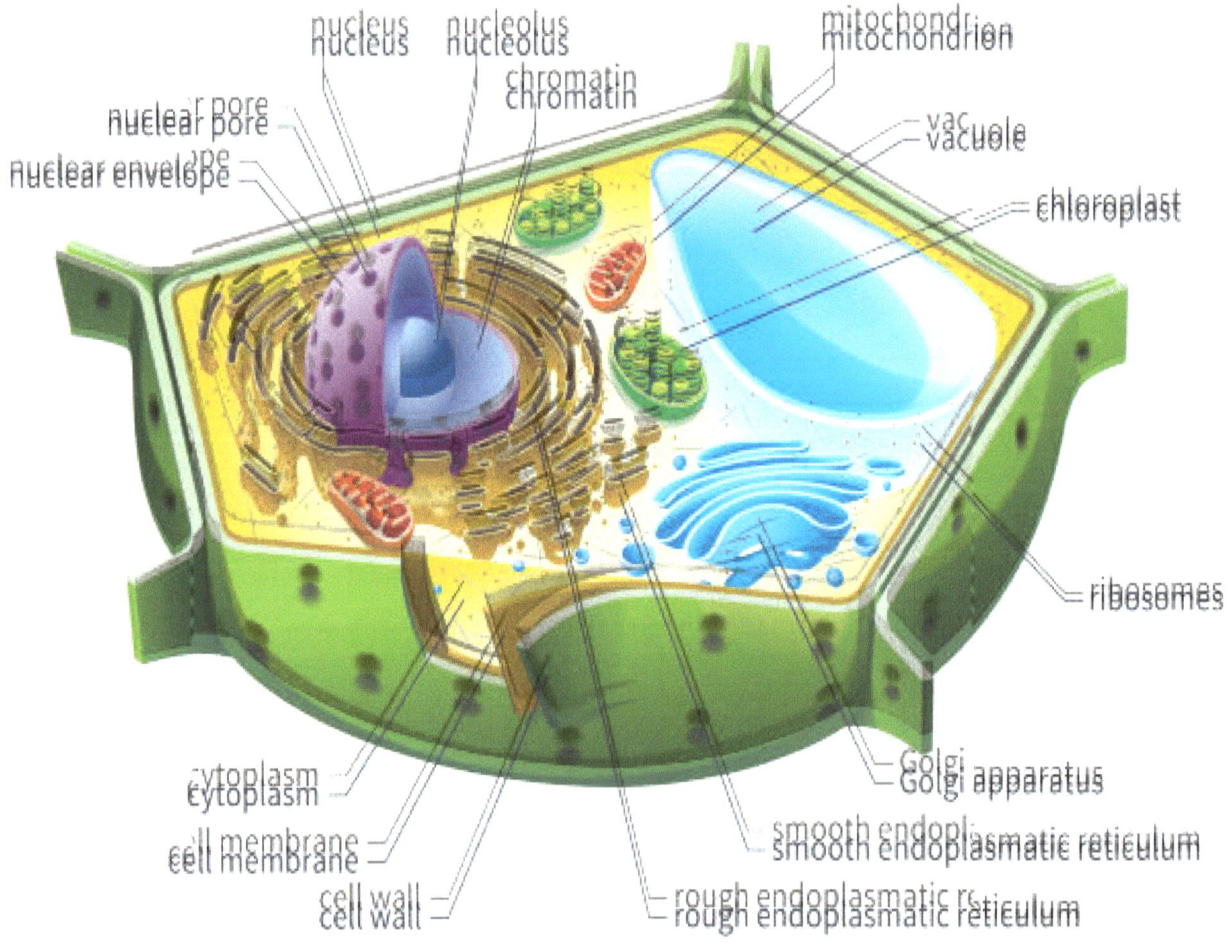

Common cells in Animals

Common cells in Humans

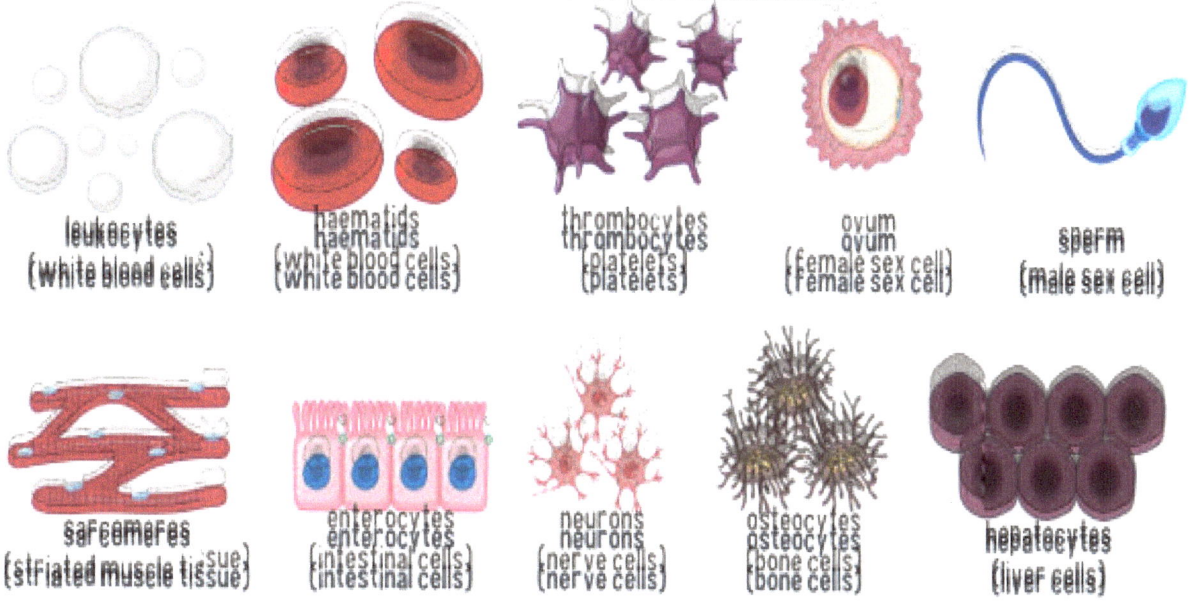

Effects of Water Pollution on Aquatic Animals

Water pollutions are absolutely harmful to many things. It's detrimental to humans, animals, as well as the environment. However, it's the aquatic animals that get affected the most. It's because they have direct contact with the pollutants. Moreover, the pollution happens on their exact habitat. In short, their habitat to which their life depends upon is being intoxicated. When that occurs, their life then is also at stake.

Just like what's already mentioned above, the most obvious of all effects of water pollution on marine animals is the disruption of their ecosystem. Oxygen reduction is certainly the number one problem caused by water pollutants. This can either be caused by excessive plant nutrients or physical pollutants. But whatever the case is, when oxygen decreases, more and more marine organisms die as well.

Moreover, mass killing of marine life is also an unfortunate known happening. The major culprit are oil spills and sewage dumping. These absolutely release poisonous chemicals that are highly intoxicating. Additionally, the other effect of water pollution in aquatic animals is the disruption of their normal growth process. That right there may not be a direct or quick impact. It's actually slow-progressing and can't be observed easily. On the other hand, it's technically more gruesome because it strips them of what's natural in them as species.

Final Thoughts

Yes, it absolutely alarming to know how the marine life is suffering from the water pollution. On the other hand, let's always bear in mind that we as humans have created the most toxic marine pollutants. For this reason, we have to solve it. It's certainly not just for ourselves but also for these aquatic animals too. We're here to take care of them and to take care of the gifts of this world in general. We definitely need to consistently step up our game to finally beat water pollution once and for all. It needs to happen because from water flows life. It's vital and that's a no-brainer to all of us. We need to solve this problem completely for a better future of our children. And our children's children. And all the generations that'll follow after them.

Water Pollution Effects on Animals: (http://gogreenacademy.com)

(photo credit: IBRRC)

Even though most people are currently unaware of the harmful effects of water pollution on animals, the issue that we have at hand is way more sinister in nature, putting at a risk the marine fauna, and the terrestrial one along with it.

While humans experience the negative effects of water pollution only when they come into contact with polluted water, animals can be more sensitive and predisposed to poisoning or perishing. The contamination of water with toxic chemicals is unfortunately irreversible, thereby resulting in a large-scale extinction of many aquatic species. Learn about the negative effects of water pollution on animals below.

In the first place, contaminating the water with industrial toxins can successfully kill many aquatic creatures, ranging from fish to frogs, tadpoles and even corals. This can further cause an important loss in food source for larger aquatic creatures, which will either perish due to the poisoned fish they feed on, or will start looking for food elsewhere. Forced to leave their natural habitat, these aquatic creatures will eventually die due to the inability to adapt to new living conditions, thus creating a chain reaction that will have only one primary result: the extinction of marine fauna.

The presence of high amounts of mercury in the water can lead to many undesirable hormonal changes in various aquatic species, which can have abnormal behavioral shifts as one of the primary outcomes. Furthermore, mercury can impact the natural and healthy development and growth of marine species, which are continually exposed to this toxic metallic chemical due to the immense industrial waste that is annually discharged in seas and oceans.

Due to the fact that industrial waste can also contain potent nutrients like nitrogen and phosphorus, toxic algae and other aquatic plants can easily grow, invading the natural habitat of many fish species, causing them to get poisoned and, in turn, to poison the animals that feed on them.

When dumping solid trash in the water, the aquatic channels are blocked, which can trap fish and other marine animals in the debris, therefore resulting in suffocating and drowning on being trapped and unable to swim. Moreover, water pollution can also affect terrestrial animals, particularly because the water coming from rivers and lakes is oftentimes used for irrigating soil and agricultural products. As a result, not only will they get contaminated, but they will also lead to many significant health issues in both animals and plants.

ANIMALS EAT PLASTIC

Animals often eat plastic because they are not always able to distinguish plastic from food. Organisms that are filter feeders (plankton, shellfish, baleen whales) or that live under the beach sand (lugworms) cannot make that distinction. Some fish eat plastic because they mistake it for fish eggs and bite at floating plastic in the water.

Turtles see plastic bags as the jellyfish that are usually on their menu. In the stomachs of the northern fulmar – which gathers its food by flying with an open beak above the water surface – plastic is almost always found. Many grazing animals on land also eat plastic. Plastic debris coated with food waste increases the chance that the plastic will be eaten.

WHAT HAPPENS WHEN ANIMALS EAT PLASTIC

Animals that accidentally eat plastic suffer and often die as a result of it. Swallowed plastic fills the stomach and not surprisingly this reduces the feeling of hunger. Animals eat less, obtain less energy, and weaken. Larger pieces of plastic can also block their gastrointestinal tract so that the plastic can no longer be excreted.

In other cases, plastic is ground into small pieces in the stomach and then scattered everywhere. In this way, the northern fulmar grinds and spreads millions of pieces every year. Some of it is left at abandoned nesting sites.

SHOCKING NUMBERS OF ANIMALS WITH PLASTIC IN THE STOMACH

Fish eat plastic. Turtles eat plastic bags. Even whales have been found dead with tons of plastic in the stomach. The stomach contents of the northern fulmar, according to long-term Dutch research, consist of an average of twenty-five pieces of plastic. A sperm whale that washed up at the Wakatobi National Park in Indonesia in December 2018 had 115 cups, 25 bags, four bottles and two slippers in its stomach.

More than a thousand pieces of plastic were counted in the whale's stomach and the total weight of plastic was six kilos. In the United Arab Emirates, plastic causes half of all camel deaths. The animals eat garbage and lumps of plastic of between ten and sixty kilos have been found in their stomachs. Because the plastic cannot pass out of the stomach, the lump continues to grow until the animal dies of starvation.

In July 2010, a young green turtle washed ashore, heavily weakened, on the coast of Brazil near Florianopolis and died a few hours later. This specimen had 3267 pieces of plastic in its gut and another 308 pieces in its stomach. Only pieces of plastic larger than 5 mm were counted.

Effects of Water Damage

Water damage is a problem that most property owners dread. When it rains heavily or Florida experiences a hurricane, the risk of water damage increases. Water can cause thousands of dollars' worth of damage to the integrity of your structure. In addition, electronics, furniture, upholstery, appliances, and other personal items may also be damaged beyond repair. Water damage also increases the risk of mold growth, which can be a very expensive problem to eliminate.

Hiring a water damage restoration company can make the cleanup process easier to handle, protect the integrity of your structure, and reduce the risk of additional problems down the road. Water Restoration Companies have workers who are experienced with damage caused by floods and storms. Qualified companies know the best ways to extract water from your home or business, remove moisture from the air and building materials, facilitate the cleanup, repair your structure, and salvage or replace damaged personal items.

Causes of Water Damage

There are several possible causes of water damage. Leaky dishwashers, clogged toilets, broken pipes, broken dishwasher hoses, overflowing washing machines, leaky roofs, plumbing leaks, foundation cracks, and hurricane and storm damage are just some of the possible causes of water damage in homes and businesses here in Central Florida.

Floods and heavy rain are other possible causes of property water damage that can lead to standing water in your home or business. Water from one of these events can result in the destruction of your property if not treated properly.

Once a home or business sustains water damage, it is important to start the water damage cleanup immediately. Starting water damage cleanup as soon as possible increases the chances that your structure, drywall and personal items can be salvaged, and decreases the chance of mold and other serious structural problems.

Assessing Water Damage

Assessing the severity of the damage should be the first step to any water damage situation. There are several different categories assigned to this type of damage. Category 1 refers to clean water, or water that does not pose a threat to humans. Possible causes of this type of damage include broken appliances or sink overflows. Category 2 water is called gray water. This means that the water is contaminated, containing microorganisms, and may cause sickness if ingested. Broken toilets, sump pumps, and seepage are common causes of Category 2 water damage. Category 3 water damage is known as black water. This type of water is dangerous, as it contains bacteria and other organisms that will cause sickness. The possible sources of black water damage include sewage problems and contamination of water left untreated for a period of time.

There are also several classes of water damage, which are important when assessing repair options. Class 1 is the least harmful form of damage. Materials have absorbed very little of the water and repair is the least involved. Class 2 has a rapid rate of evaporation, which means that carpets and furniture cushions may be damaged. Water damage repair is more difficult when Class 2 damage is involved. Class 3 Water Damage has the fastest rate of evaporation. In this case, the water may come from broken sprinklers or other overhead sources, soaking the walls and furniture. This type of damage involves walls, ceilings, insulation, and personal items. All of these classes of damage are best handled by a water damage restoration company to avoid the possibility of mold and other problems. However, Class 4 damage requires special water restoration and water removal procedures, making it necessary to involve professionals. This type of damage may affect hardwood floors, plaster, and concrete.

Most restoration companies will provide an assessment for free, determining the category of water and type of damage involved. To protect the integrity of your structure, the value of your property, and the health of your family, it is important to contact a water damage specialist to professionally assess any water damage.

The Restoration Process

The water restoration process is critical. The right procedures and equipment can save precious belongings and even prevent property from becoming condemned. Water restoration companies specialize in minimizing the effects of water damage. The success for water damage restoration depends on accurate and complete upfront assessment and proper treatment of the affected areas with the right equipment. Restoration companies use high-tech equipment and well-documented procedures to remove flood water and minimize damage. Many water restoration companies only handle water removal, not the complete restoration of your home. One Restore and other full-service water restoration companies also offer repair and reconstruction. Be sure to ask before hiring a restoration contractor what services they provide.

Restoration Contractor Addresses Health Hazards

It is important to hire a restoration company as soon as possible after water damage occurs. Moisture promotes the growth of mold and other organisms, increasing the risk for serious damage to your property and potential health problems. Mold exposure may aggravate allergy and asthma symptoms, especially in children and people with compromised immune systems. Exposure to mold may also increase the risk for respiratory diseases and other medical problems. Proper and immediate cleanup of water damaged areas can help avoid mold and related health problems.

Throughout this book it will include,
some needed stops of interlude:
Allowing you to link and think
of all you've read and all you've viewed:
So many things you've taken in,
a pause because SO MUCH ACCRUED!
Let's take a breath, get more in depth
to learn of more you may have known:
a common THOUGHT that really OUGHT
to CHANGE with MORE that's shown!
Yes, water's a necessity for all of us, both you and me.
We often take SO MUCH for granted,
at times INDEED need reprimanded!
Yet, we ARE learning: churning and yearning
for what's at stake for global sake...
and SOMETIMES in need a comedy break!

Avoiding NOT to soil it: Often, running for a toilet!

When needing a restroom
we simply CAN'T presume:
that this MUCH NEEDED RUSH surely WILL flush!

Ways and facts to act: Knowing more, we onward grow
into a global based society that cares for ALL humanity.
Little things that we CAN do, things within OUR POWER
help to glean and clean our world... each AND every hour.

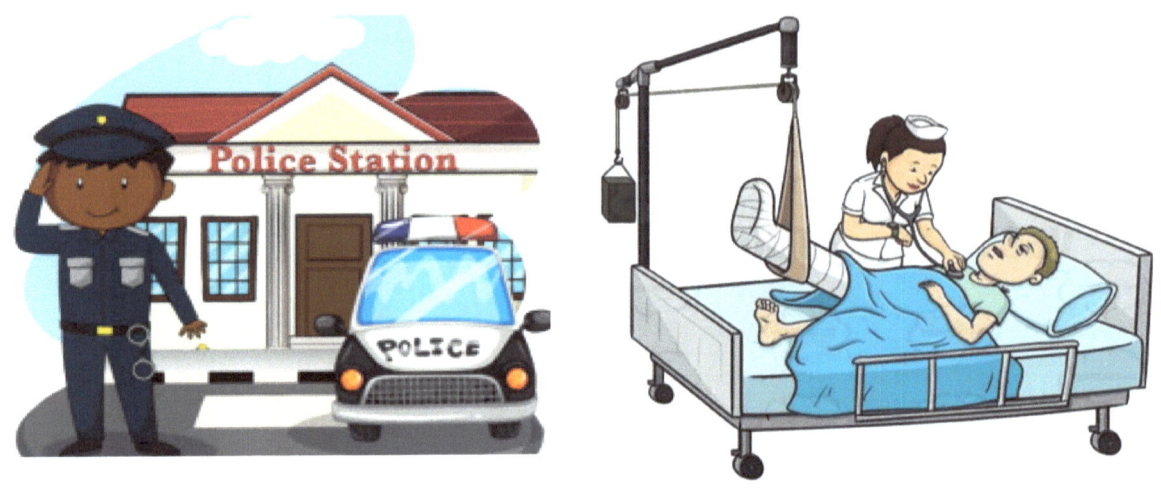

National AND Community services are needing our help-

environmental TRAGEDIES are making them yelp!

The fire hydrants we commonly see
get water from the ocean's seas.
There's SO much MORE fire crews do: Let's NOT conclude
but let's pursue to know MORE than we THOUGHT we knew!

Firefighters all aspire: They do FAR MORE than put out fire. Let's find out! Get to the core! Onward go! Let's read some more, see what's in store: LET'S GO EXPLORE!

The storm produced gusty winds, which contributed to power outages in parts of the Southeast and mid-Atlantic, Weather.com said. Almost 850,000 homes and businesses had lost electricity in Virginia, Georgia, Tennessee and the Carolinas as of midday Monday, according to poweroutagre.com, but power was slowly restored throughout the day with about 500,000 still without power as of 6p.m.

Snow falls at the White House on Jan. 3, 2022.

In the mid-Atlantic, the heavy, wet snow accumulated on power lines, leading to outages, the Weather Service said. A cold front associated with the storm will be the focus for scattered showers and thunderstorms across the Carolinas and into Florida on Monday. Damaging winds and a few tornadoes are the main concerns in this severe weather threat.

Ever wondered WHY fire trucks and teams often advance with an ambulance at circumstance urgency-emergency scenes?

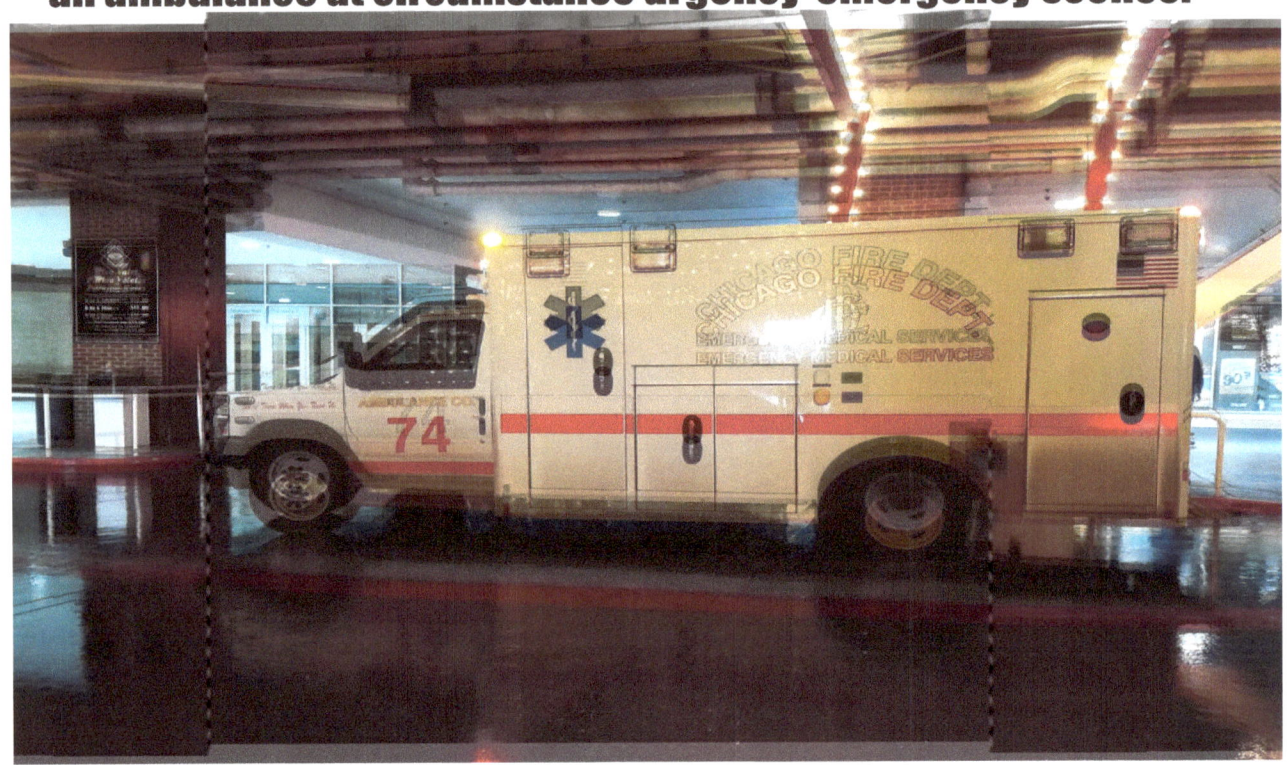

Every 23 seconds a fire erupts in the United States, according to the National Fire Protection Association, and, in 2011 alone, there were over 1.3 million fires that required fire department intervention. Not only are firefighters brave, they are also very busy!

Firefighters often begin their careers by volunteering at local firehouses, joining the junior firefighter league or undergoing training through municipal fire companies. They might also choose to pursue a program in fire science, a path that could end with an associate or bachelor's degree. The following resources are meant to help aspiring firefighters understand what the job requires, plan for the education and training they will undergo and keep abreast of the field with the latest research, publications and blogs from firefighters who are in the line of duty every day.

Why Send A Firetruck To Do An Ambulance's Job?

April 11, 2017 4:23 AM ET Heard on Morning Edition: Chicago's Fire Department has only half as many ambulances as it has fire vehicles, but it gets 20 times more medical calls than fire calls these days. Monica Eng/WBEZ. Let's say you think you're having a stroke and you call 911 for an ambulance. In a lot of cities across the country there's a good chance that a firetruck — with a full fire crew including a paramedic — will race to your door. But that doesn't mean they can deliver the emergency care you might need. In Chicago, like many cities, the fire department oversees both firefighters and paramedics who work on ambulances.

A Chicago Fire Department ambulance and firetruck respond to a call in 2016. *Arvell Dorsey Jr./Flickr*

"They save valuable time by sending the closest vehicle, which is usually a firetruck that has at least one paramedic and a lot of equipment on it," Langford says. When a medical call comes in, dispatchers usually prefer to send ambulances. But there are half as many of those as there are firetrucks. And fire department spokesman Larry Langford says those ambulances can be super busy: "They'll hit the street at 7 o'clock in the morning and may not come back to the firehouse at all until 7 o'clock the next morning," he says. Meanwhile, firetrucks are often much less busy fighting fires.

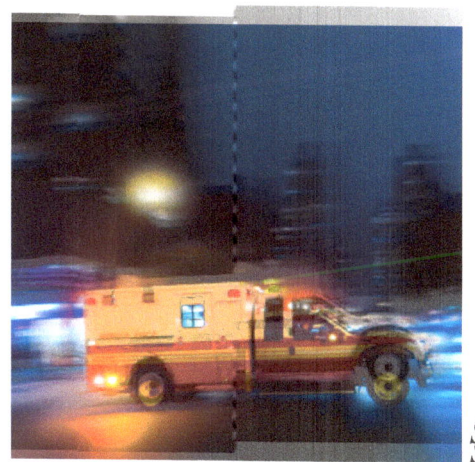

SHOTS-HEALTH NEWS: *Paramedic Shares His Wild Ride Treating 'A Thousand Naked Strangers'*

That sounds logical until you ask why Chicago's fire department still has twice as many firetrucks as ambulances, especially when the department gets 20 times more medical calls than fire calls. Getting answers can be difficult. That has a lot to do with the political power of fire departments and their unions and the challenge of trying to change that

SHOTS-HEALTH NEWS Replacing An Ambulance With A Station Wagon

Chicago's not alone in facing these challenges. Most cities are seeing big drops in fire calls and big jumps in medical calls. But few are really reforming their departments to meet this changing emergency landscape. Portland State researcher Phil Keisling thinks that's a mistake. He looked at why fire departments don't just admit that they're mostly medical services these days. «And I keep getting answers that are really not a whole lot more than, ‹Well, that›s the way we›ve always done it,› » he says. Keisling says that›s not a good answer «in a world that has limited resources and you want to try to optimize the resources you›ve got.»

As more cities see the drawbacks of using giant firetrucks for medical issues, they're facing calls for reform. That's what Misty Bruckner found when she researched the problem at the Public Policy Center at Wichita State. While she didn't find agreement on everything, she said there was some consensus.

SHOTS-HEALTH NEWS Doctors Make House Calls On Tablets Carried By Houston Firefighters

"I think everybody can agree that the ladder truck responding to someone who may have a sprained ankle is not the best use of our public resources," she says. Langford disagrees. And he thinks people shouldn't get so hung up on what kind of vehicle arrives. He's even got a catch phrase for it: "Don't look at the conveyance. Look at the care."

Catchy or not, the conveyance can matter. Firetrucks aren't equipped to take you to the hospital. Only ambulances are. And this transport part can be crucial, according to veteran Chicago paramedic Rich Raney. "When you get a stroke patient or a trauma patient, the most important thing is that they be transported to the hospital as quickly as possible," he says. "As they say with stroke patients, time is brain, basically."

Each city runs its emergency services differently, so solutions are going to vary. In Chicago, for example, paramedics want more ambulances and staffing. New York and Wichita recently started deploying medics in SUVs for less urgent calls. And Washington, D.C., is trying something called nurse triage lines. They let callers talk through their problems with a nurse on the phone. But Keisling says some proposals should also look at moving resources from large firefighting staffs. "And it's not anti-firefighter, it's not anti-union, and it's not anti-government," he says. "It's just, why aren't we taking limited resources and deploying them in a smarter way?"

While there's no agreement on exactly what that smarter way is going to be, most agree it doesn't involve sending a firetruck to treat someone with heartburn.

(Monica Eng *is a reporter with NPR member station WBEZ. You can follow her* @monicaeng)

Why do fire departments require paramedics?
Since most fire departments run upwards of 70% or more emergency medical-related responses, it only makes sense for them to require this as a prerequisite. Additionally, it is much less training they have to provide you during the recruit academy. Many departments are looking for licensed paramedics.

What can help prevent homes and businesses catching on fire more often?
It turns out that it can be attributed to a few pretty mundane factors: stricter fire codes, fireproof building materials, cars that catch on fire less often, and installation of protective devices like smoke alarms.

Fire crews are needed and serve so many ways you can explore, you'll be surprised by the resources that help maintain this core:
Firefighting, medical, technology, ecology, research, personal interviews with those who serve the public. (see *reference pg.63)*

At times, we need to laugh at ourselves at what we do.
Other times, must realize, internalize the harm we do: BOTH me and you!
The hope of our world is in our hands to save our water, air and lands.
Of course, WE'RE THE SOURCE of change: THE GAUGE!
To guide our stride just turn the page.

Water Waste should be Erased!

Identify all the things / activities that are harmful to the environment in the image below

**Fighters for the world FROM ALL THE HARM that's hurled,
SAVING WATER CREWS pursue pollution and abuse:
Defending to the end: This awful wasteful trend,
Together we CAN MEND = ON US IT DOES DEPEND!**

LOOKING into the WHYS
helps us to devise
and THUS SO to be wise...
prompting us to do
something MORE than criticize!
Studies done to UNDERSTAND
keeps us ALERT to STOP the HURT
of DAMAGE done- to TAKE COMMAND!
They say, "It takes a village to raise a child."
I say: It takes a "call to all" to stop NOT stall

THIS 'RILED WILD'
EARTH
DEFILED!

(Please review references not previously noted within: pgs.61-62)

ABOVE IMAGE COPIED FROM BOOK 3

Scientists and all who care, have studied, written and have shared!
Start studies with buddies, friends trending for ending:
Explore some more- these references: Simply DON'T make inferences.
Discern and learn! Get the facts to act within these reference pages:
Let's start RIGHT NOW and DO our part: Disengaging these outrages
by disarming the **ALARMING, HARMING**
RAGES OF WASTED WATER STAGES!

HYDRO AND FRACKING

Mom Junction's health articles are written after analyzing various scientific reports and assertions from expert authors and institutions.

Project Oceanography; University of South Florida
Types of Water Pollution; University of North Carolina at Chapel Hill
Summer Science Lesson Plan: Water Pollution; Harvard University
Environmental Health and Economic Impacts of Road Salt; New Hampshire Department of Environmental Services. Cheevaporn & P. Menasveta, Water pollution and habitat degradation in the Gulf of Thailand; National Center for Biotechnology Information
S.S. Lang, Water, air and soil pollution causes 40 percent of deaths worldwide, Cornell research survey finds, Cornell University
Water pollution is on the rise globally; UNESCO
8. The Effects: Human Health; United States Environmental Protection Agency
9. R. Singh et al.; Heavy metals and living systems: An overview; National Center for Biotechnology Information
10. The Effects: Economy; United States Environmental Protection Agency
11. Why is pH = 7 the neutral point?; University of Illinois
12. What is the biggest source of pollution in the ocean?; National Oceanic and Atmospheric Administration
13. Water Facts – Worldwide Water Supply; US Department of Interior – Bureau of reclamation

WATER POLLUTANTS

Alrumman SA, El-kott AF, Kehsk MA. Water pollution: Source and treatment. American journal of Environmental Engineering. 2016;6(3):88-98. (2) Briggs D. Environmental pollution and the global burden of disease. British medical bulletin. 2003;68:1-24. (3) Bibi S, Khan RL, Nazir R, et al. Heavy metals in drinking water of LakkiMarwat District, KPK, Pakistan. World applied sciences journal. 2016;34(1):15-19. (4) Khan N, Hussain ST, Saboor A, et al. Physiochemical investigation of the drinking water sources from Mardan, Khyber Pakhtunkhwa, Pakistan. International journal of physical sciences. 2013;8(33):1661-71. (5) Pawari MJ, Gawande S. Ground water pollution & its consequence. International journal of engineering research and general science. 2015;3(4):773-76. (6) Juneja T, Chauhdary A. Assessment of water quality and its effect on the health of residents of Jhunjhunu district, Rajasthan: A cross sectional study. Journal of public health and epidemiology. 2013;5(4):186-91. (7) Khan MA, GhouriAM. Environmental Pollution: Its effects on life and its remedies. Journal of arts, science and commerce. 2011;2(2):276-85. (8) Owa FD. Water pollution: sources, effects, control and management. Mediterranean journal of social sciences. 2013;4(8):65-8. (9) Kamble SM. Water pollution and public health issues in Kolhapur city in Maharashtra. International journal of scientific and research publications. 2014;4(1):1-6. (10) Ho YC, Show KY, Guo XX, et al. Industrial discharge and their affects to the environment. Industrial waste, In Tech. 2012:1-32.
11. Desai N, Smt Vanitaben. A study on the water pollution based on the environmental problem. Indian Journal of Research. 2014;3(12):95-96. (12) Jabeen SQ, Mehmood S, Tariq B, et al. Health impact caused by poor water and sanitation in district Abbottabad. J Ayub Med Coll Abbottabad. 2011;23(1):47-50. (13) Yonglong Lu, Song S, Wang R, et al. Impacts of soil and water pollution on food safety and health risks in China. Environment International. 2015;77:5-15. (14) Khurana I, Sen R. Drinking water quality in rural India: Issues and approaches-Water Aid. India water Portal. 2008: (15) Ebenstein AY. Water pollution and digestive cancer in China. Institutions and governance programs. 2008:1-45. (16) http://research.gsd.harvard.edu/hapi/
(17) Halder JN, Islam MN. Water pollution and its impact on the human health. Journal of environment

and human. 2015;2(1):36-46. (18) Ahmad SM, Yusafzai F, Bari T, et al. Assessment of heavy metals in surface water of River Panjkora. Dir Lower, KPK Pakistan. J Bio and Env Sci. 2014;5: 144-52. (19) Corcoran E, Nellemann C, Baker E, et al. Sick water? The central role of wastewater management in sustainable development. A Rapid Response Assessment. United Nations Environment Programme. 2010.

(20) Nel LH, Markotter W. New and emerging waterborne infectious diseases. Encyclopedia of life support system. 2009;1:1-10. 21. Ullah S, Javed MW, Shafique M, et al. An integrated approach for quality assessment of drinking water using GIS: A case study of Lower Dir. Journal of Himalayan Earth Sciences. 2014;47(2):163-74. (22) Krishnan S, Indu R. Groundwater contamination in India: Discussing physical processes, health and sociobehavioral dimensions. IWMI-Tata, Water Policy Research Programmes, Anand, India. 2006. (23) Currie J, Joshua GZ, Katherine M, et al. Something in the water: contaminated drinking water and infant health. Canadian journal of economics. 2013;46(3): 791-810. (24) Ahmed T, Scholz F, Al-Faraj W, et al. Water-related impacts of climate change on agriculture and subsequently on public health: A review for generalists with particular reference to Pakistan. International journal of environmental research and public health. 2013;13:1-16. (25) Salem HM, Eweida EA, Farag A. Heavy metals in drinking water and their environmental impact on human health. ICEHM. 2000;542-56. (26) Chowdhury S, Annabelle K, Klaus FZ. Arsenic contamination of drinking water and mental health. 2015;1-28. (27) https://www.nps.gov (28) http://www.in.gov/isdh/22963.htm (29) Ballester F, Sunyer J. Challenges to public health in the new millennium. Journal Epidemiol Community Health. 2000; 54:2-3. 3p. Andersson I, Fenger BH. Environment and human health. European environment agency. 2003

[50 Fantastic Fire Science Resources - Fire Science](#) and volunteer sources

[Water Cycle | Biology | Ecology - Bing video](#) results in multiple links with options for various grade levels and age groups

[https://youtu.be/0Puv0Pss33M](#) How to Save Our Planet ages 10-adult

[Ecosystems: The Water Cycle: StudyJams! Science | Scholastic.com](#) 4th grade and up Presentation video

[Natural water cycle game | South East Water Education (educationsoutheastwater.com.au)](#) Video game ages K-5 grade plus

PG. 70(A) NEIN! NO! LA! HET! NON! NEE! NOPE! NO in different languages "indicates" the sound in various languages: NEIN (German "nine"), NO (Spanish, Italian, "noh" English) NON (French "noh"), LA (Arabic "lah"), HET (Russian "Nyeht") NEE (Dutch "nay" same in Swedish but written NEJ "nay") NOPE (English when saying NO with mouth closed!) [Do You Know How to Say No in Different Languages?](#)

[https://www.npr.org/2020/05/22/858800112/herd-like-movement-of-fuzzy-green-glacier-mice-baffles-scientists](#) (First discovered in 1950: Herd Of Fuzzy Green 'Glacier Mice' Baffles Scientists, May 22, 2020)
Look this up and you will see: A PHENOMENAL ANOMALY that's been found:
This video shows with sight and sound what's actually growing on glacier ground!
Copy, paste online with haste! Listen and read! Yes, indeed! You've GOT TO SEE!
Check ON, come ON, LINK ON: This truly UNCOMMON PHENOMENON!

Finally, let's turn the page for helpful tips
to SNIP NOT SKIP….STOP water drips!
Let's ALL make haste; SAVE WATER WASTE

 Water your yard and outdoor plants early or late in the day to reduce evaporation.

 Use a shut-off nozzle on your hose.

 Use plants that require less water.

Mulch around plants to hold water in the soil.

 Get an Energy Star labeled washing machine. Wash only full loads.

Use a low flow showerhead.

 Take shorter showers — five minutes or less is best.

 Turn off the water while soaping hands and brushing teeth.

Turn off sink faucet while scrubbing dishes and pots.

 Install new toilets that use less than 1.6 gallons per flush.

  Put faucet aerators on sink faucets.

 Us a broom, not a hose, to clean driveways and walkways.

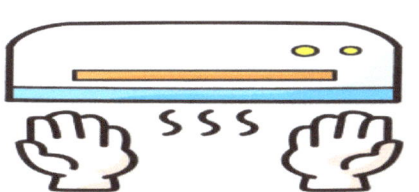

**Electric hand dryers INSTEAD of paper towels, helps and SAVES:
One TON of paper destroying 17 trees!
Making three cubic yards of LANDFILL GRAVES:**

**7,000 gallons of water *POLLUTED* NEEDED and CRAVED,
this zapping KEEPS happening, IT gets MORE DEPRAVED!**

**Another THING that *STINGS* that NEEDS indeed
to CHANGE with SPEED! Hand cleaning signs NEED to read:**

FIRST SOAP HANDS!

**NOT "FIRST run water, NEXT use soap"
NEIN! NO! LA! HET! NON! NAY! NOPE!** (A)

**Wasting by FIRST running water ~ we NOW know:
WE SHOULD NOT OUGHT TO!**

Rosey & Roughest
support and agree that we must save world energy.
UP to YOU and UP to ME... IT'S UP TO ALL, YES, GLOBALLY!

www.ingramcontent.com/pod-product-compliance
Lightning Source LLC
Chambersburg PA
CBHW051954210526
45473CB00029B/861